Gretchen's Guide

to

Weeds and Wild Flowers

on

Key School's 8th Grade

Wye Island Trip

by

Gretchen Nyland

Dedicated to the 8th graders who learned to survive the togetherness. Make it happen.

ISBN: Softcover 978-1-5144-8342-8
EBook 978-1-5144-8341-1

Print information available on the last page

Rev. date: 06/13/2016

To order additional copies of this book, contact:
Xlibris
1-888-795-4274
www.Xlibris.com
Orders@Xlibris.com

The trip that everyone remembers

The Leaders

Lee Curry, the English teacher, lacrosse coach who was the original environmental leader at Key School, others have followed filling his shoes with much success; the 8th grade dedicated science teachers searching for birds above and insects below; our esteemed Latin teacher, teaching canoeing and being the water gatherer; Jim G. (now a screenwriter, playwright, graphic novel creator) returning for the weekend with his skulls and bones to hunt for footprints and more bones; the other Lee , the ever ready helper; and the many graduates returning to help; and me, the ex-art teacher who messes around with identifying and using wild plants, appearing every year.

The Students

The 8th grade, 1976 - now

My involvement

I figure I won't be able to go on these annual trips many more times, in fact, I may have gone on my last one after 40 years of participating. Too much effort to sleep in a tent. I've taken the list of plants from those seen over the last 40 years and picked the plants that appeared almost every year on our 2 hour walks. There are more and different plants each year, but due to different weather patterns, things disappear and can reappear the following year.

The students gather in their groups and I inform my group to get their notebooks and make 4 columns. (I have not included trees in this booklet)

1.-Trees

2.-vines

3.-bushes

4.-small weeds or wild flowers

Can everyone identify poison ivy? And with that we're off to see if we can identify all those many other green leaves. Walking along the tent line we can see poison ivy, wineberries, blackberries, trumpet vines, honeysuckle, wild roses, and Virginia creeper growing entwined all through the under brush and into the trees. Grape vines are appearing, and low down there are Indian strawberries and wild onions. Trees such as; oak, sassafras, walnut, cedar, Osage orange, maple, ash, and sweet gum, which they usually mistake for maple, make the canopy. As we go down the driveway the berries and the vines take over. Curly dock appears as do more wild rose bushes. There are good examples of sumac and elderberry growing together, both with their compound leaves. At the road a patch of yarrow appears with its feathery leaves and white umbrella-like flowers. We stop here and smell the crushed leaves that make an herbal tea and listen to my stories. We stop for some shade as it is usually hot, and we identify some red and white clovers that show up here. Who knew they grew so differently and could be poisonous if fermenting?

Across the road one can find crane's bill, both broad leaf and narrow leaf plantains, and cleavers. And with more sunlight available along the road there are sheep and wood sorrels, both best known as sour grass. The three blue flowered plants; Virginia day flower, chicory, and blue-eyed grass usually show up along the road. Also shepherd's purse and pepper grass appear along the edge. Every few years we will see the small purple Venus's-looking-glass. There are now big stands of poke weed and milkweed sticking up above all the rest of the growth and easy to identify. In the parking lot area there are more elderberry bushes with their white blossoms and red and white mulberry trees from which we can do some eating if the weather was right for them to produce. (some years the berries are still too green). We walk past stands of milkweed , where we pick some for dinner) and a cherry tree identified by the small still green drupes on our way back to the Wye Holly, and finally a large grouping of jewelweed, known as a poison ivy cure with its beautiful bluish juicy stems. The Holly tree was hit by lightening before our 2015 trip. Not climbable any more. Very sad, but our herbaceous friends mostly still come through.

Poison Ivy (*Rhus radicans)*

Vine:

Leaflets - (3), compound, alternate

Can everyone identify poison ivy? The first question, the first plant that we identify with 3 leaves (There are more) and one that we don't want to touch. The word *urushiol* is tossed around and pronounced with difficulty. This is the poison oil that you don't want to get on you. Everyone writes it in their notebook.

Bittersweet *(Celastrus orbiculatus)*

Vine:

Leaves - simple, opposite

Not easy for us to spot in early June, but very invasive as it develops. Later on it develops green fruits (poison) turning yellow and splitting open with red-orange seeds in the fall. Often used in floral decorations.

Japanese Honeysuckle (*Lonicera japon)*

Vine:

Leaves - opposite

Flowers - yellows and whites:

Yes, you can suck a drop of nectar from the end of a honeysuckle flower. Question: Which is best –the yellow or the white? There are lots of this vine behind the tent row.

Left	Right
Common Blackberry	**Wineberry**
(Rubus allegheniensis)	*(Rhubus phoenicolasis)*
Bush:	Bush:
Leaves - 3 toothed leaflets	3 toothed leaflets
Flowers - white, 5 petals	white, 5 petals
Canes - greenish, thick sharp thorns	reddish, recurved bristles
Fruit - black	red

Very easy to identify. The wineberries have their red fuzzy looking canes, although the berries are not ripe until July. The blackberries are forming small green berries that will be ripe in August. This is Brer Rabbit's bush that he hid in.

Virginia Creeper (*Parthenosissus quinquifolia)*

Vine:

Leaves - 5, compound

Berries - blue black, may have enough oxalic acid that will harm kidneys.

Can you tell the difference between this and poison ivy? It used to grow on the walls at school, but it was taken down once they decided that it wasn't poison ivy and they could touch it.

Indian Strawberry *(Duchesnea indica)*

Vine:

Leaflets - 3

Flowers - 5 petals, yellow

Berries - red and tasteless

Found around school and good to draw, but just mush to eat. Found in mowed areas along driveway.

Wild Grape *(Vitis* riparia)

Vine: with tendrils

Leaves - simple, alternate

Berries - small, purple cluster, but haven't developed yet when we are there in June.

Always a disappointment that there are no grapes, just the remains of the flower. Leaves used in Greek cooking.

Trumpet Vine *(Campsis radicans)*

Vine: climbs high

Leaves - compound, opposite

Flowers - orange, tubular

Seeds - long pod like

Climbing with other vines (poison ivy, grape, Virginia creeper) up one tree. Hard to figure out which is which. You can put the flowers on each finger to play witch's hands.

Both called Plantain – broadleaf and narrow-leaf

Left - *Plantago major*

Right - *Plantago lanceolatel*

Common low weed plants growing in basal rosettes along the drive. All veins come from center making them "monocots" like corn. Can be used when young in soups. Narrow-leaf is good for shooting heads.

Sumac *(Rhus copallina)* *left*

Bush:

Leaves - compound, alternate

Berries - drooping, rusty red, hanging clusters

If there were more bushes, we could make Indian lemonade like we did in 3rd grade when we made dyes, but I know of only one bush. Hopefully more will sprout. It grows next to the elderberry bush and the two are confusing to identify.

Elderberry *(Sanbucus canadensis)* right

Bush:

Leaves - compound, opposite

Flowers - white, flat-topped:

Berries - not developed, but will be small and purple.

One can return and collect berries in August and make jelly, pie or wine. The pith is soft and you can hollow it out to make whistles or stems for your clay pipes like the Colonists did..

Wild Rose ***(Rosa multiflora)***

Vine or bush:

Leaves - alternate, 5-7 compound leaflets

Flowers - white, 5 petals, good smell

Thorns - curved, barbed, and fierce

All along the tent line and down the driveway. Most of the flowers are gone, and the green berries have begun to form when we get there. These and the blackberry thorns hurt the most.

Curly Dock *(Rumex cripus)*

Small plant: to 4 ft. tall

Leaves - curly edges because veins curl back in, alternate, long and narrow

Seeds - dense, turning tan and becoming rusty red .

Easy to back-pull seeds off. Each small seed has one drop of carbohydrate. When the seeds have turned rusty, one can winnow them to get the husks off, but for us in June they are not ripe enough. (and it's too much trouble). Often used in seed bouquets.

Red Clover *(Trifolium pretense) top - pink*

Small plant:

Leaves - 3 leaflets from flower stalk, white chevron design.

Flowers - purple to pink flower head.

White Clover *(*Trifolium repens*)* bottom - white

Small plant:

Leaves - 3 leaflets from underground stolons, not stalked

Flowers - white flower head:

One can see them together at the road. The red ones grow tall and the white ones are small .

One can make a tea from fresh leaves and flowers, but it has been found that fermenting clover leaves produce coumarin which gets transformed and stops blood from clotting. Cows that have eaten fermenting clover have died from internal hemorrhages.

Cranesbill or Wild Geranium *(Geranium maculatum)*

Small plant:

Leaves – deeply lobed

Flowers – 5 petals, pink

Seeds - elongated, capsule.

Sometimes hard to find along the road. When found, one can take 2 capsules, split one with thumb nail, insert the second one through slit and you have a little scissors. Hm-m-m, too childish for them?

Common Day Flower *(Commelina communis)*

Small plant:

Leaves - simple, forms sheath at base around flower and stem

Flowers - 2 blue petals, 1 white

Beautiful blue

Can eat both leaves and flowers

Common weed in gardens

Yarrow *(Achilleae millefolium)*

Small plant:

Leaves - alternate, fern-like

Flowers - small, composite, umbrella-like

Good for a bitter tea.

Named after Achilles who used it to cure his soldier's wounds, but my story is that his mom held him by his heel and dipped him into the yarrow juice where he eventually got shot, thus "Achilles heel". The 8th graders know *mille* = thousand and *folium* means leaf and they learn the Latin name easily.. It's also known that if you put one of the feathery leaves under your pillow, the first one you see in the morning will be your true Love. 8th graders laugh at that one as there are many sleeping in each tent.

Left	Right
Sheep Sorrel	**Wood Sorrel or Sour Grass**
(Rumex acetosella)	*(Oxalis stricta)*
Small plant:	Small plant:
Leaves - alternate, arrowhead	3 heart-shaped leaflets
Flowers - cluster of tiny reddish	5 yellow petals

Leaves of both are good to eat and very sour, but both contain oxalates so do not eat too much. Wood sorrel develops beans that young kids like to find and eat. You can spot sheep sorrel in lawns in spring by the rusty looking patches in the grass. The leaves of the sour grass look like hearts as opposed to clover leaves.

Left	Right
Shepherd's Purse	**Wild Peppergrass**
(Capsella bursa-pastoris)	*(Lepidium virginicum)*
Small plant:	small plant:
Leaves - basal rosette	deeply toothed, often stalked
Seeds - flat, heart-shaped	flat, round

Both have seeds and leaves that are good to eat - pungent and peppery and no poisonous look-alikes.

Pokeweed *(Phytolacca americana) or* **InkBerry**

Small plant:

Leaves - elliptical, tapering at both ends

Flowers - long racemes with small white flowers

Fruit - drooping clusters of dark purple-black berries

Tall plant. alternate leaves, small white flowers in racemes that develop green berries that will turn deep purple.

Can be eaten only when small (less than 8 inches), as purple coming up the stem from the root is a poison and causes it to be not edible.

Berries can be used for ink. George Washington used them.

Blue-eyed grass *(Sisyrinchium ?)*

Small plant:

Leaves - grass-like, thin

Flowers - blue, sometimes more violet

Another beautiful blue flowered

Sometimes hard to spot

Clevers or Bedstraw *(Galium aparine).*

Small plant: sprawling with backward hooked soft bristles

Leaves - long in whorls

Flowers - small, white

Good for a game of throwing at each other so it sticks on one or the other.

Chicory (*Cichorium intybus)*

Small plant:

Leaves - rosette of deeply toothed basal leaves

Flowers - sky-blue rays

Roasted roots good for substitute coffee

Beautiful blue flower - open when It's sunny

Found along sunny road .

Milkweed (Aesclepius *syriaca)*

Small plant:

Leaves - opposite, toothless, thick

Flowers - umbrella –like clusters, pink-purple

Seed pods - warty, small cucumber looking, filled with silky seeds

Love of the Monarch Butterfly

Bitter, white, sticky, latex when picked

Good to eat new buds, new leaves, new pods, when prepared right

BUT, we get just a taste because the monarchs need it more than we do.

Complicated flower structure when you really look at it

Jewelweed, Touch-me-not, or **Wild Impatiens** ***(Impatiens biflora)***

Small plant:

Leaves - lower leaves are opposite, long, oval, toothed

Flowers - sacked shape with spur, yellow-orange with spots

Seeds - bean shape that explodes the seeds when touched

Last plant for us to identify on the trail

Poison ivy cure

One can eat new growth

Hollow, transparent, juicy plant

Touch-me-not seeds fly all over when touched

Makes a yellow-orange dye

Bibliography

Brill, S. 1994 *Identifying and Harvesting Edible and Medicinal Plants in Wild (and not so wild) Places. New York:* HarperCollins

Del Tredici, P. 2010 *Wild Urban Plants of the Northeast: a field guide. New York:* Cornell University Press

Clements, S. and C. Gracie. 2006. *Wildflowers in the Field and Forest: A Field Guide to the North-eastern United States. New York: Oxford University Press*

Eastman, John. 2003 *The Book of Field and Roadside.* Mechanicsburg, Pa.: Stackpole Books

Jones, Pamela.1991 *Just Weeds History, Myths, and Uses.* New York: Prentice Hall Press

Newcomb, L., 1977 *Newcomb's Wildflower Guide,* New York: Little Brown

Peterson, T. R. and Margaret McKenny.1968. *Peterson Field Guides -Wildflowers Northeastern/ North-central North America.* New York: Houghton Mifflin.

Thayer, S. 2006 *The Forager's Harvest: A Guide to Identifying, Harvesting, and Preparing Edible Wild Plants.* Birchwood, Wi.: Forager's Harvest.

Thieret, John W., 2001 *National Audubon Society-A Field Guide to the North American Wildflowers Eastern Region. Revised Edition.* New York: Chanticleer Press, Inc.

Taught at the school for forty-one years. Hoping to keep the interest in art and in the environment alive and well, figuring the more you know, the better off you will be.

www.ingramcontent.com/pod-product-compliance
Lightning Source LLC
Chambersburg PA
CBHW050810180526
45159CB00004B/1618
9781514483428